Grade 2

©2003 Options Publishing Inc.
All rights reserved. No part of this document may be reproduced or used in any form or by any means — graphic, electronic, or mechanical, including photocopying, recording, taping, and information storage and retrieval systems.

Acknowledgments
Product Development: Margaret Fetty
Editor: Michelle Howell
Cover Illustration: Tad Herr
Design and Production: Creative Pages, Inc.
Production Supervision: Sandy Batista

ISBN 1-59137-114-7
Options Publishing Inc.
P.O. Box 1749
Merrimack, NH 03054-1749
www.optionspublishing.com
Phone: 800-782-7300 Fax: 866-424-4056

All rights reserved. Printed in USA.
15 14 13 12 11 10 9

Dear Parent,

   Summer is a time for relaxing and having fun. It can also be a time for learning. *Summer Counts!* can help improve your child's understanding of important skills learned in the past school year while preparing him or her for the year ahead.

   Summer Counts! provides grade-appropriate practice in subjects such as reading, language arts, vocabulary, and math. The ten theme-related chapters motivate your child to continue learning throughout the summer months.

   When working through the book, encourage your child to share his or her learning with you. With *Summer Counts!* your child will discover that learning is a year-round process.

Apreciados padre,

   El verano es una época para descansar y divertirse. También puede ser una época para aprender. *Summer Counts!* puede ayudar a que su hijo(a) mejore las destrezas importantes que aprendió el pasado año escolar al mismo tiempo que lo(a) prepara para el año que se aproxima.

   Summer Counts! provee la práctica apropiada para cada grado en las asignaturas como la lectura, las artes del lenguaje y las matemáticas. Los diez capítulos temáticos incluyen actividades y rompecabezas que motivarán a su hijo(a) durante el verano.

   Cuando trabaje con el libro, anime a su hijo(a) a que comparta lo que ha aprendido con Ud. Si Ud. desea puede desprender la página de las respuestas que aparece en la parte trasera del libro. Puede usar la misma para revisar el progreso de su hijo(a). ¡Con *Summer Counts!* su hijo(a) descubrirá que el aprendizaje puede ocurrir en cualquier momento—inclusive en el verano!

Table of Contents

Letter 2

Chapter 1: Flags
The Stars and Stripes 4
Nouns: People, Places, and Things 6
Rhyming Words 7
Counting Whole Numbers 8
Even and Odd Numbers 9
Flag Facts 10
My Flag 11

Chapter 2: Parks
In the Park 12
Nouns with Capital Letters 14
Opposites 15
Number Places 16
Number Fun 17
Swing Scramble 18
The Perfect Park 19

Chapter 3: Animals
Two Silly Mice 20
Verbs 22
Rhyming Words 23
Dollars and Cents 24
How Much? 25
Animal Word Search 26
Animal Mix-Up 27

Chapter 4: Plants
Growing Beans 28
Verbs: Singular and Plural 30
Watching Words Grow 31
Less or Greater 32
All About Temperature 33
Tree Talk 34
Flower Power 35

Chapter 5: Neighborhoods
What Is a Neighborhood? 36
Verbs: Past Tense 38
Neighborhood Ride 39
Comparing Costs 40
Graph It 41
My Neighborhood 42
My Favorite Place 43

Chapter 6: Rain
Make Your Own Rainbow 44
Verbs: Now or Past 46
Rhyming Words 47
Rounding Numbers 48
Time for Rain! 49
Rainy Day Puzzle 50
All Wet 51

Chapter 7: Travel
From Engine to Caboose 52
Nouns and Pronouns 54
Opposite Ways 55
Rounding Money 56
Packing Problems 57
Travel Codes 58
On the Move 59

Chapter 8: Space
Stars Fill the Sky 60
Adjectives 62
Rhyming Words 63
Make an Estimate 64
Spinning in Space 65
Space Word Search 66
Creature Features 67

Chapter 9: Tricks
Spider Tricks Snake 68
Telling Sentences 70
The Same Game 71
Adding: Carry the One 72
Twenty-Pound Package 73
Tricky Riddles 74
What a Trick! 75

Chapter 10: Cooking
The Gingerbread Man 76
Ask a Question 78
Kitchen Connection 79
Subtract: Find the Difference 80
Lines of Symmetry 81
Food Scramble 82
Favorite Food Fun 83

Reading Check-Up 84
Math Check-Up 88
Certificate of Completion 91
Answer Key 93

CHAPTER 1

The Stars and Stripes

Did you know that our flag has a nickname? It is "the Stars and Stripes." The flag has that name because it has white stars and red-and-white stripes on it. The very first flag of America had 13 stars and 13 stripes on it. The stars and stripes stood for the first 13 states. Today, our country has 50 states. Our flag has one star for every one of those 50 states.

Our flag stands for us, the people. It stands for where we live in our 50 states. The flag stands for our towns, too. This is why "the Stars and Stripes" waves over places like the post office and the firehouse.

On the Fourth of July, you see lots of flags. People march with "the Stars and Stripes." People wave flags from their houses. On the Fourth of July, we think about what "the Stars and Stripes" stands for. It stands for us!

The Stars and Stripes

Directions Use what you have just read to answer the questions.

1. What would be another good title for this story?
   _____

2. How many stars are on the flag?
   _____

3. What do the stars stand for?
   _____

4. In what places can you see an American flag? Name two.
   _____
   _____

5. Is this selection real or make-believe? How do you know?

Reading Comprehension *Grade 2*

Nouns: People, Places, and Things

REMEMBER

Naming words are called **nouns**. Some nouns name people. Other nouns name places or things.

EXAMPLES

The **girl** runs. (person)
Lila walks to the **park**. (place)
Carl has a **hat**. (thing)

Name the Nouns

Directions Circle the noun in each sentence. Then tell if the noun names a person, place, or thing.

1. Ana rides in a parade. _____

2. She has a flag. _____

3. Her friends wave. _____

4. Ana rides by the school. _____

5. The parade stopped at the park. _____

6. It went all through town. _____

Language *Grade 2*

Rhyming Words

Directions Draw lines to match the words in the first row to the rhyming words in the second row.

1. pool   class

2. pass   flag

3. bag   school

4. sends   friends

Ben's First Day

Directions Read the story. Use words from the box below to fill in the blanks.

| class | flag | school |
| teacher | friends | |

Ben was scared. It was his first day in a new

(5) _____. His sister helped him find his first

(6) _____. The (7) _____

smiled and showed Ben to his desk. Then the children

in the class stood up. They faced the (8) _____

and said the pledge to the flag. Everyone wanted to meet

Ben. By lunch, Ben had made two new (9) _____.

Vocabulary Grade 2

Counting Whole Numbers

REMEMBER

We count **whole numbers**. Whole numbers can be written as **digits**. Whole numbers can also be written as **words**.

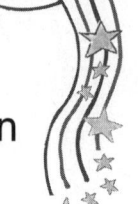

EXAMPLES

digits   words
0, 1, 2, 3   zero, one, two, three

Write It Out

Directions Write each number in words.

1. 5 _____
2. 2 _____
3. 7 _____
4. 4 _____
5. 10 _____
6. 9 _____
7. 1 _____
8. 6 _____
9. 8 _____
10. 3 _____

Misplaced Numbers

Directions Write the missing numbers for each line.

11. 1, 2, _____, 4, _____, 6, 7, _____, _____, 10
12. 2, 4, _____, 8, _____, 12, _____, 16, _____, _____
13. 5, 10, _____, _____, 25, _____, 35, 40, _____, _____
14. 10, _____, 30, _____, 50, _____, 70, _____, _____, 100

8 Math *Grade 2*

© 2003 Options Publishing Inc. No copying permitted.

Even and Odd Numbers

REMEMBER

An **even number** ends in 0, 2, 4, 6, or 8. An **odd number** ends in 1, 3, 5, 7, or 9.

Even or Odd?

Directions Write if the number is <u>even</u> or <u>odd</u> on the line.

1. 7 _____
2. 4 _____
3. 19 _____
4. 32 _____
5. 58 _____
6. 99 _____

Directions Write if each group of stars is even or odd.

7.
   _____

8.
   _____

9.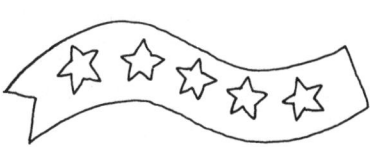
   _____

10.
    _____

Math Grade 2

Flag Facts

Directions Read each definition. Use words from the box below to finish the puzzle.

> blue   stars   country   white
> flag   states   stripes

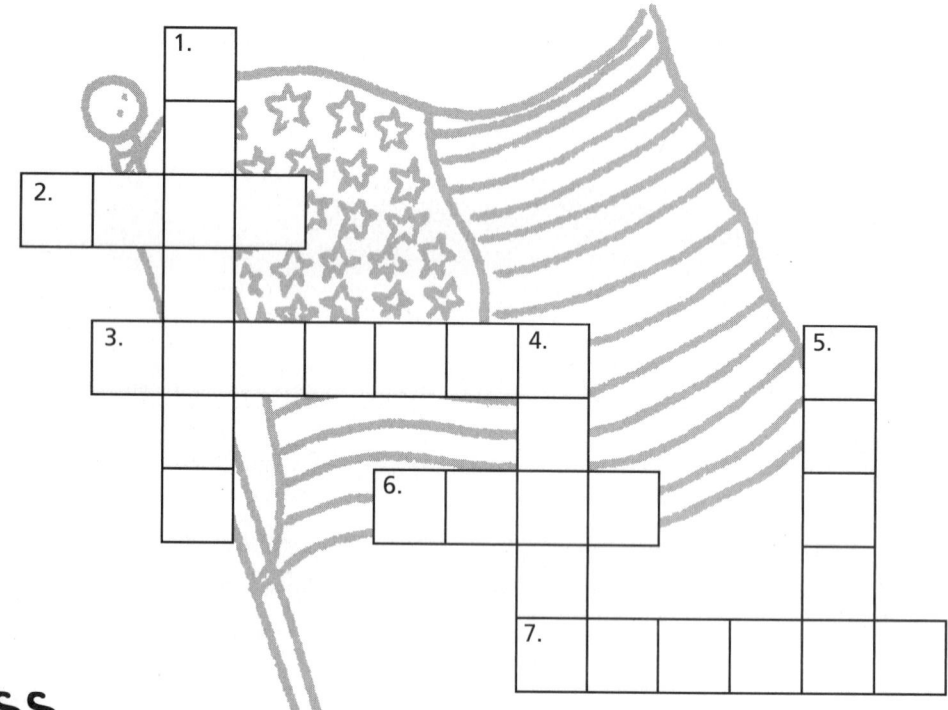

ACROSS

2. Color on the American flag that is the same color as the ocean
3. Lines on the flag are also called this.
6. A piece of colored cloth that stands for a country
7. America has 50 of these.

DOWN

1. A place where people live together with the same rules
4. There are 50 of these on the American flag.
5. The color of the stars on the American flag

Crossword Puzzle *Grade 2*

My Flag

Directions The American flag stands for the American people. Draw a flag that would stand for you. The flag could show pictures of things you like to do. Write sentences that tell what the pictures and colors on your flag mean.

Drawing/Writing *Grade 2*

CHAPTER 2

IN THE PARK

It was a hot, sunny day. Lucy's dad packed their basket. Lucy and her dad began walking to the neighborhood park. They stopped on the way to get Lucy's friend, Jake.

At the park, Lucy picked a spot under a tree. "Let's eat here," she said. Dad pulled sandwiches and fruit out of the basket. Everyone enjoyed the picnic. After lunch, Lucy and Jake fed the birds. Then Dad tossed a ball to Lucy and Jake. Everyone smiled and laughed.

"Can we swing now?" asked Jake.

"Let's pick up our trash first," said Lucy's dad. "The only thing we should leave in the park is our footprints."

"I know," said Lucy. "The only thing we should take with us is pictures. Would you take our picture, Dad?"

It was a great picture!

In the Park

Directions Use what you have just read to answer the questions.

1. Where does this story take place?
   _____

2. Name two things Lucy, Jake, and Dad did.
   _____
   _____

3. Did everyone have a good time in the park? How do you know?
   _____
   _____

4. Lucy's dad said, "The only thing we should leave in the park is our footprints." What does he mean?
   _____
   _____

5. Number the following sentences 1, 2, or 3 in the order that they happened.

   _____ After lunch, Lucy and Jake fed the birds.

   _____ "Let's pick up our trash first," said Lucy's dad.

   _____ Lucy and her dad began walking to the neighborhood park.

Reading Comprehension *Grade 2*   **13**

Nouns with Capital Letters

REMEMBER

A **noun** names a person, place, or thing. Some nouns name special people, places, or things. These special nouns begin with capital letters.

EXAMPLES

	Nouns	Special Nouns
Person	girl	Maria
Place	town	Brownwood
Thing	holiday	Thanksgiving

Correct Capitals

Directions Underline the nouns in each sentence that name a special person, place, or thing. Then write each sentence correctly.

1. We live in elmwood.

   _____

2. My favorite park is on beech street.

   _____

3. My friend alex lives near the park.

   _____

4. We meet at the park on saturday.

   _____

Language Grade 2

Opposites

Directions Read each word. Find the word from the box below that is opposite in meaning. Write it on the line.

| after | love | outside | sky |
| cloudy | stay | take | sit |

1. inside _____ **2.** go _____

3. before _____ **4.** give _____

5. hate _____ **6.** stand _____

7. ground _____ **8.** sunny _____

Puppy's Poem

Directions Read the poem. Use words from the box below to fill in the blanks.

| park | run | outside | side |

I take my puppy to the **(9)** _____.

She loves to **(10)** _____ and play outside.

We stay **(11)** _____ until it's almost dark.

Then she walks home by my **(12)** _____.

Vocabulary Grade 2

Number Places

REMEMBER
Place value tells the value of a digit.

EXAMPLE
257 has three digits

2 is in the **hundreds** place.
5 is in the **tens** place.
7 is in the **ones** place.

hundreds	tens	ones
100s	10s	1s
2	5	7

What's the Value?

Directions Write the value of the underlined digit.

1. 3̲8 _____
2. 2̲17 _____
3. 45̲ _____
4. 16̲7 _____
5. 2̲95 _____

Write It Out

Directions Write the number for each number word.

6. one hundred, six tens, and four ones _____
7. three hundreds, one ten, and five ones _____
8. six hundreds, two tens, and eight ones _____

16 Math Grade 2

© 2003 Options Publishing Inc. No copying permitted.

Number Fun

Directions Use the cards below to answer the questions.

1. What is the **smallest** number that can be made with 2 cards? _____

2. What is the **smallest** number that can be made with all 3 cards? _____

3. What is the **largest** number that can be made with 2 cards? _____

4. What is the **largest** number that can be made with all 3 cards? _____

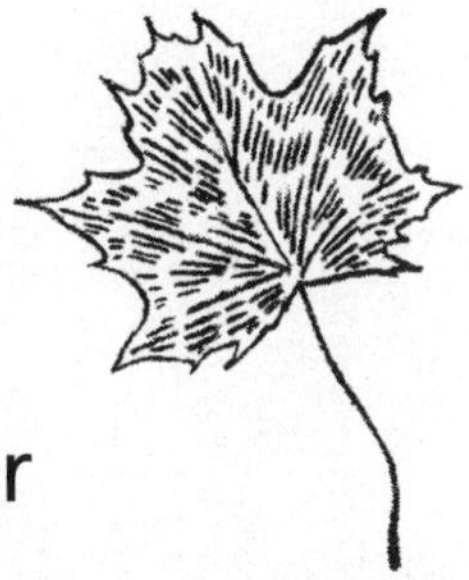

5. Use the cards to make a number that is larger than 450 but smaller than 600. _____

6. Write 4 numbers you can make that are smaller than 50.

_____ _____ _____ _____

Math Grade 2

Swing Scramble

Directions Each clue names something fun about the park. Read the clues. Then unscramble the letters to find the words. Write the words on the lines.

1. erca To run really fast __ __ __ __
                              1

2. istoude Not inside __ __ __ __ __ __ __
                           2

3. luhag A happy sound __ __ __ __ __
                          3   5

4. gniws Something on a playground that moves forward and backward __ __ __ __ __
                                                                              4

Directions Use the numbered letters to solve the riddle. Write the letters on the lines.

What did the dog say when he walked on rocks?

__ __ __ __ __   __ __ __ __ __
1 2 3 4 5   1 2 3 4 5

18 Word Scramble Grade 2

The Perfect Park

Directions What do you like to do at the park? Are there special games you like to play? Draw a picture of the perfect park. Write sentences that tell what you would do at the perfect park.

Drawing/Writing *Grade 2* 19

CHAPTER 3

Two Silly Mice

One day, two mice went for a walk. They were hungry. They saw some cheese. The mice knew they had to share. They wanted the pieces to be the same size. Each mouse was afraid that the other would get a bigger piece. They needed to split the cheese between themselves. The mice did not know how to do this. They talked about it. The mice became angry.

Just then a cat came by. She said that she could help. The cat had a plan. She broke the cheese in two. She made sure one piece was bigger. The mice were not happy about this. The cat said she could fix it. She took a big bite out of one piece of cheese. Now the other piece of the cheese was bigger.

The cat kept taking bites out of the cheese. This went on and on until the cat could eat no more. She left two pieces for the mice. They were just the same size. They were tiny!

Two Silly Mice

Directions Use what you have just read to answer the questions.

1. What is this selection about?
   _____

2. Why did the mice want pieces of cheese that were the same size?
   _____

3. What clue told you that the cat was going to play a trick?
   _____
   _____

4. How do you think that the mice felt at the end of the story? Why?
   _____
   _____

5. Is this selection real or make-believe? Tell how you know.
   _____
   _____
   _____

Reading Comprehension *Grade 2*

Verbs

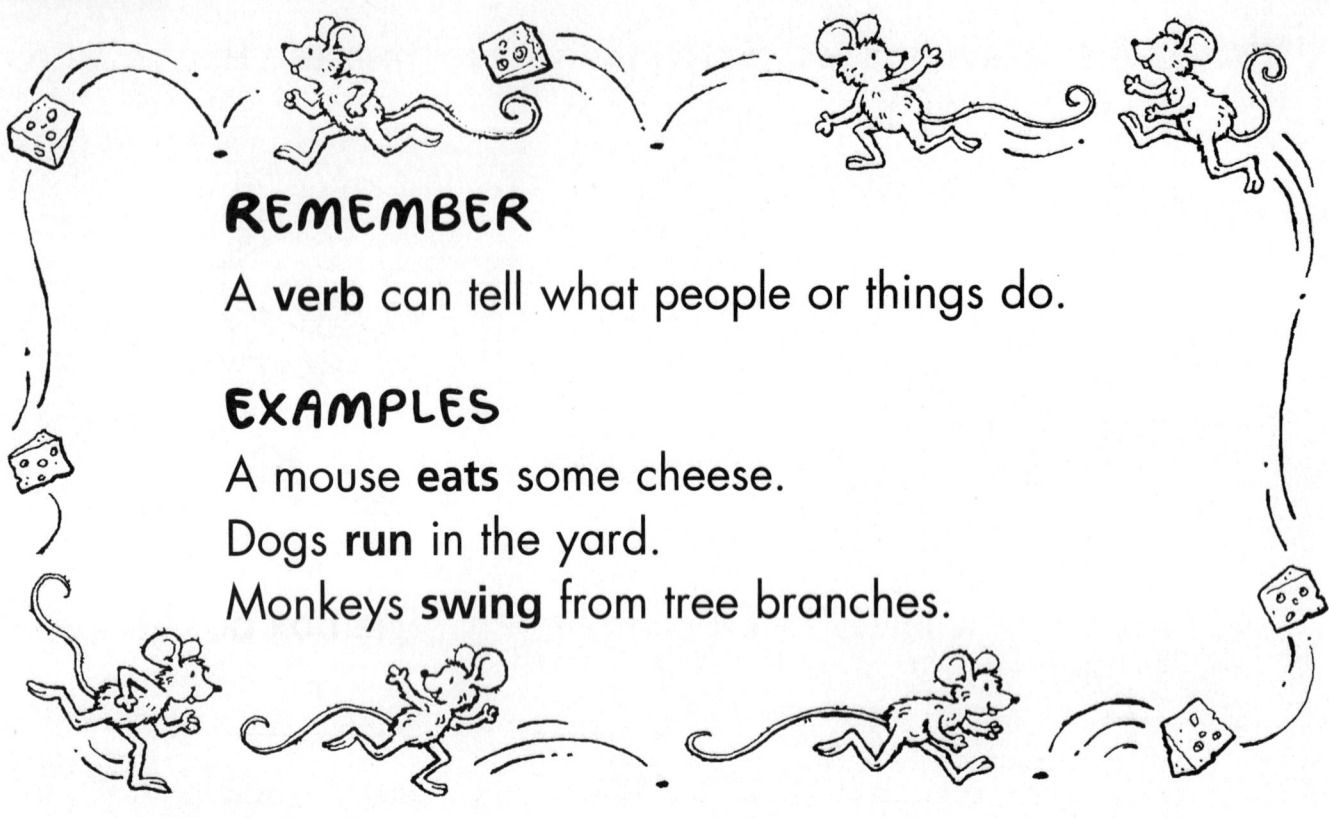

REMEMBER

A **verb** can tell what people or things do.

EXAMPLES

A mouse **eats** some cheese.
Dogs **run** in the yard.
Monkeys **swing** from tree branches.

Animal Actions

Directions Circle the verb in each sentence.

1. Bees live in a hive.

2. The owl opens its eyes.

3. An elephant swings its trunk.

4. Fireflies glow at night.

5. Birds sing in the trees.

22 Language Grade 2

Rhyming Words

Directions Draw lines to match the words from the first row to the rhyming words in the second row.

1. keep   dear
2. arm   saw
3. ear   farm
4. claw   sheep

Farm Finds

Directions Read the letter. Use words from the box below to fill in the blanks.

| farm | many | goat | milk | saw | very |

Dear Ella,

We are on a **(5)** _____. There are **(6)** _____ sheep here. I even **(7)** _____ a horse. It was **(8)** _____ big. The **(9)** _____ is funny. It likes to eat cheese. We watched the farmer **(10)** _____ a cow.

Your friend,
Daniel

Vocabulary Grade 2

Dollars and Cents

REMEMBER

Dollars and **cents** are written with special symbols.

EXAMPLE

dollar sign → $4.25 ← decimal point
                ↑   ↑
             dollars cents

Cents can be written using only the cent sign (¢). No dollar sign is needed.

EXAMPLE

15 cents   15¢

Cents Sense

Directions Write how much money the pictures show.

1.

   2 quarters, 1 dime,
   1 nickel, 2 pennies

2.

   1 dollar, 1 quarter,
   1 nickel, 4 pennies

_____   _____

The Cost of Cheese

Directions The price of each piece of cheese is given in words. Write the price in dollars and cents on the tags.

3.

   forty-nine cents

4.

   one dollar and
   twenty-four cents

24 Math *Grade 2*

How Much?

Directions Sam went to the store to buy some cheese. Look at the sign to help Sam answer the questions.

1. What is the price for 1 piece of cheese? _____

2. What is the price for 3 pieces of cheese? _____

3. Sam has 6 quarters, 3 dimes, and a nickel. How much money does Sam have? Write the amount in both words and with money signs. _____
_____

4. What coins will Sam give the clerk to buy three pieces of cheese? _____
_____

5. How much money will Sam have left? Write the amount in both words and with money signs. _____
_____

6. What can Sam buy with the money he has left? Circle it.

Math Grade 2 25

Animal Word Search

Directions Find the words listed in the box. Then circle them in the puzzle. The words are hidden across and down.

cow	cat	duck	horse
dog	goat	sheep	mouse

t	g	o	a	t	m	i	l
m	i	t	h	o	r	s	e
o	d	u	c	k	n	d	d
u	l	s	p	s	c	o	w
s	t	h	g	u	a	g	a
e	s	e	c	a	u	w	r
c	k	e	a	c	g	e	k
d	z	p	t	s	h	u	p

26 Word Search *Grade 2*

Animal Mix-Up

Directions Use the body parts of different animals to draw a make-believe animal. You could draw an animal with an elephant's trunk and the feet of a duck. Then write sentences telling where the animal lives and what it eats.

Drawing/Writing *Grade 2*

CHAPTER 4
Growing Beans

WHAT YOU NEED

bean seeds   jar with lid

paper towels   dirt

water   pot

WHAT TO DO

1. Get 2 or 3 bean seeds.

2. Wet some paper towels.

3. Put the paper towels in a jar.

4. Put the seeds on top of the paper towels.

5. Close the lid. Let the seeds stay there for several days. You will see sprouts.

6. Put dirt in a pot.

7. Put the sprouted beans in the dirt.

8. Water the beans and put the pot in the sun.

Watch your bean plants grow!

Growing Beans

Directions Use what you have just read to answer the questions.

1. Read the steps below.

    Write 1, 2, 3, or 4 to show the order.

    _____ Close the lid.

    _____ Wet some paper towels.

    _____ Put the paper towels in a jar.

    _____ Put the seeds on top of the paper towels.

2. Why should steps for growing a seed be done in order?
   _____

3. What are two things you need to give your plants to help them grow? _____

4. What is a sprout? _____
   _____

5. Draw three steps that show how to plant a sprout.

Reading Comprehension Grade 2

Verbs: Singular and Plural

REMEMBER

A **verb** can show action that is happening now. An **-s** is added to the end of a verb if the word tells about one.

EXAMPLES

One student **plants** tomato plants. (This sentence tells about one student.)

All the students **plant** tomato plants. (This sentence tells about more than one student.)

Time to Plant

Directions Circle the correct verb in each sentence.

1. Fran (ask, asks) her friends to come over.

2. She (want, wants) help planting seeds.

3. Mark and Juan (plant, plants) carrot seeds.

4. They (water, waters) the seeds.

5. Mia (look, looks) for bean seeds.

6. Soon green plants (grow, grows) from the seeds.

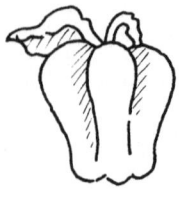

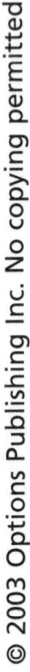

Watching Words Grow

Directions Read each word. Add or take away letters to make the words listed in the box below. Write the new words on the lines.

any	grow	seed	our
tall	than	tree	plant

1. many _____
2. all _____
3. four _____
4. row _____
5. street _____
6. tan _____
7. plan _____
8. see _____

A Pretty City

Directions Read the story. Use words from the box above to fill in the blanks.

You can grow almost **(9)** _____ kind of plant in the city. Buy seeds and **(10)** _____ them in the ground at your local park. They will **(11)** _____ very fast. You can also plant a **(12)** _____. It will grow big and **(13)** _____. It will grow much slower **(14)** _____ the flowers. This is what makes **(15)** _____ parks beautiful!

Vocabulary Grade 2   31

Less or Greater

REMEMBER
When you compare numbers, you find which number is **less than** or **greater than** the other. Look at the digits in the highest place value and compare them. Use > and <.   > = greater than   < = less than

EXAMPLES

Which is greater?
30   25
Compare 3 and 2.
30 > 25 because 3 > 2.

Which is less?
127   140
Both begin with one, so compare 2 and 4.
127 < 140 because 2 < 4.

More or Less

Directions Look at each set of numbers. Write < or > between each pair showing which one is less or greater.

1. 30 _____ 40
2. 22 _____ 20
3. 12 _____ 14
4. 86 _____ 89

What's The Address?

Directions The map below shows seven houses on Rosebud Circle. Write the addresses from the list.
Hint: Addresses get larger as you go.

5.

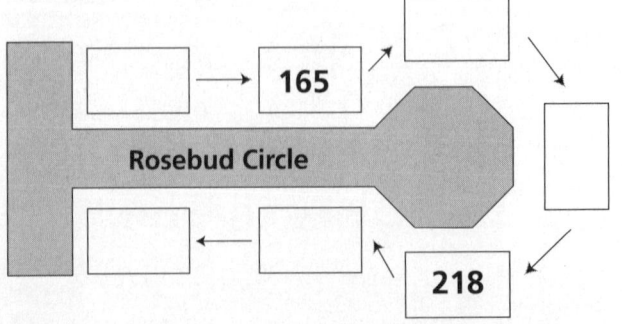

Address List
174
278
128
200
242

32 Math Grade 2

All About Temperature

REMEMBER

The **temperature** tells how hot or cold a place is. A **thermometer** measures temperature. Temperature is measured in **degrees**. We use **Fahrenheit degrees** (°F) to measure temperature.

EXAMPLE

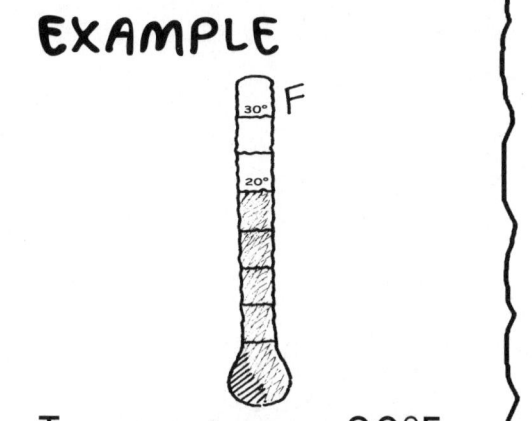

Temperature = 20°F

Weather Watch

Directions Use the thermometer to answer questions 1 and 2. Then complete the page.

1. What temperature is shown on the thermometer?

    Temperature = _____ °F

2. What if the temperature drops 20 degrees? What will the thermometer show?

    Temperature = _____ °F

3. Shade each thermometer to show the temperature.

    a. 40°F   **b.** 75°F   **c.** 60°F

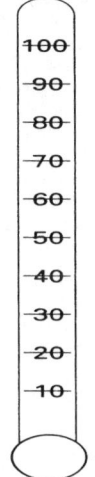
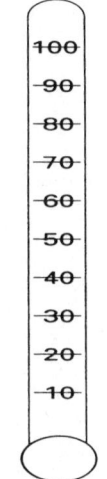
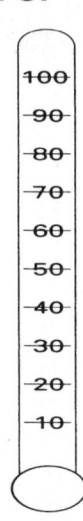

Math Grade 2   33

Tree Talk

Directions The answer for each riddle is written in code. Each number stands for a letter. Use the key to solve the code. Write the letters on the blanks to answer the riddles.

A	B	C	D	E	F	G	H	I	J
1	2	3	4	5	6	7	8	9	10

K	L	M	N	O	P	Q	R	S	T
11	12	13	14	15	16	17	18	19	20

U	V	W	X	Y	Z
21	22	23	24	25	26

1. What did the tree say when the deer ate some of its leaves?

   __ __ __ __ __ __ __ __ __ __ __.
   12 5 1 6 13 5   1 12 15 14 5

2. How did the little tree feel when he grew beautiful, red berries?

   __ __ __ __ __ __ __ __ __ __
   9 20   23 1 19   2 5 18 18 25

   __ __ __ __ __.
   8 1 16 16 25

3. How can you tell a dogwood tree from an oak tree?

   __ __ __ __ __ __ __ __ __.
   2 25   9 20 19   2 1 18 11

4. What did the little tree say when it started to grow?

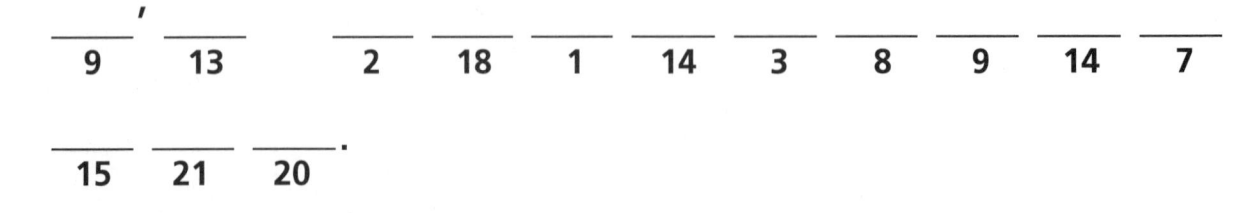

   __ , __ __ __ __ __ __ __ __ __ __
   9   13 2 18 1 14 3 8 9 14 7

   __ __ __.
   15 21 20

34 Code Fun Grade 2

Flower Power

Directions Many flowers have special meanings. A yellow rose may mean **friendship**. A red rose may mean **love**. If you grew a new kind of flower, what color would it be? What would it mean? Draw a picture of the flower. Then write sentences telling about the flower.

CHAPTER 5

What Is a Neighborhood?

What is a neighborhood? It is a place where people live, work, and play together. Many families live in a neighborhood. People might live in an apartment or a house. They might work in a shop selling things or in the post office taking in the mail. They might play at a park.

Some neighborhoods are busy. They have many people going places. The buildings might be tall and close together. Other neighborhoods are more quiet. They may have houses and yards. You can see the homes lined up in rows. Each neighborhood has its own special places. But what is the same about every neighborhood? Neighbors live there!

What Is a Neighborhood?

Directions Use what you have just read to answer the questions.

1. What is a neighborhood?

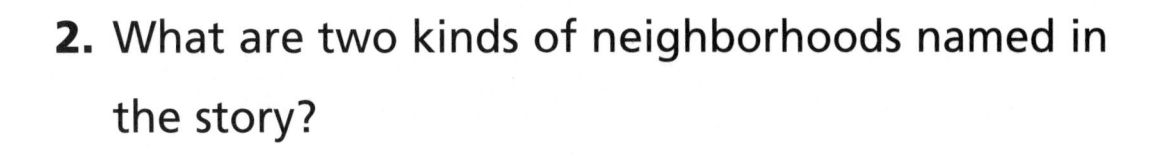

   _____

2. What are two kinds of neighborhoods named in the story?

   _____
   _____

3. What kind of neighborhood do you live in?

   _____
   _____

4. How might a farm be different from a neighborhood?

   _____
   _____

5. Which of these sentences below is a fact? Circle it.

   My neighborhood is the best.

   People live in a neighborhood.

Reading Comprehension *Grade 2*

Verbs: Past Tense

REMEMBER

A **verb** can show action that happened in the past. The letters **-ed** are added to the end of a verb if the word tells about an action that happened in the past.

EXAMPLES

Angie **packed** her suitcase on Monday.
She **visited** her aunt last weekend.

A Neighborhood Walk

Directions Circle the correct verb for each sentence.

1. Last summer, Angie (visits, visited) her aunt.

2. On the visit, Angie and her aunt (walk, walked) around the neighborhood every day.

3. Angie (learned, learns) about the places in the neighborhood that summer.

4. She (talks, talked) to the people she met.

5. Angie (enjoyed, enjoys) her visit last summer.

6. Angie was happy when her aunt (asks, asked) her to come back.

Neighborhood Ride

Directions Draw a line from the word on the left to its clue.

1. bike   not on

2. fix   to make something work right

3. fell   it has two wheels

4. off   tripped over

The Bike Ride

Directions Read the story. Use words from the box below to fill in the blanks.

bike	fell	thank
fix	chain	riding

I was **(5)** _____ my

(6) _____ in my neighborhood yesterday.

I rode over a bump, and I **(7)** _____ over.

The **(8)** _____ came off my bike! I stopped

and tried to **(9)** _____ it. A neighbor was

walking down the street. "Can I help?" he asked. "Yes,

(10) _____ you," I said.

"I cannot fix it by myself."

Vocabulary Grade 2

Comparing Costs

REMEMBER
When you compare the cost of things, you find which is the better buy. Looking at the dollars and cents will help you compare prices. Use the signs > and <, too.

EXAMPLES

Which is greater?
39¢   47¢
Compare 3 and 4.
39 < 47 because 3 < 4.

Which is less?
$1.42   1.29
Each has one dollar, so compare 4 and 2.
$1.42 > 1.29 because 4 > 2.

Money Matters

Directions Read each question. Write < or > between each pair of prices showing how the prices compare.

1. 19¢ _____ 24¢

2. 34¢ _____ 28¢

3. 41¢ _____ 45¢

4. $1.13 _____ $2.09

5. $1.58 _____ $1.85

6. $1.67 _____ $1.60

The Better Buy

Directions Look at each tag. Which would be the best buy, or the lowest price? Circle your answer.

7. 49¢   58¢

8. $1.44 $1.28

Graph It

Directions Some children sold lemonade and snacks to buy new books for the library. The graph below shows how much money they raised. Use the graph to answer the questions.

Money Raised for Library

- $7.20 Fig Bars
- $9.45 Lemonade
- $6.75 Popcorn
- $4.50 Apples

1. What type of graph is shown? Circle your answer.

    bar graph   circle graph

2. How many kinds of foods did the children sell? _____

3. Each item costs 45¢. Which food did the children sell the most of?

    _____

4. Which food did the children sell the least of?

    _____

5. How much more money did the children make selling lemonade than fig bars? Show your work.

    _____

Math Grade 2

My Neighborhood

Directions The words below name places in a neighborhood. Draw a picture of what each place looks like in your neighborhood.

1. My school

2. The library

3. A friend's house

4. A store

My Favorite Place

Directions What is your favorite place to go in your neighborhood? Write about why this is your favorite place to visit.

CHAPTER 6
Make Your Own Rainbow

Have you seen a rainbow? We mostly see them after it rains. What makes a rainbow?

Light shines through tiny drops of water in the air. Then colors show. The colors of the rainbow are red, orange, yellow, green, blue, and purple.

You can make your own rainbow. Get a garden hose. Go outside on a bright day. It is best to go before noon. Stand with the sun behind you. Spray the water into the air. Make fine drops of water. Wave the hose around until you see your own rainbow.

Make Your Own Rainbow

Directions Use what you have just read to answer the questions.

1. Read the steps below. Write **1**, **2**, or **3** to show the order.

   _____ Wave the hose around.

   _____ Spray fine drops of water into the air.

   _____ Stand with the sun behind you.

2. When do people mostly see rainbows?
   _____

3. What makes a rainbow?
   _____

4. Name three colors in a rainbow.
   _____
   _____

5. Is this selection real or make-believe? How do you know?

   _____
   _____
   _____
   _____
   _____

Reading Comprehension *Grade 2*

Verbs: Now or Past

REMEMBER
The verb **be** is special. **Am**, **is**, and **are** tell about things that are happening now. **Was** and **were** tell about things that happened in the past.

EXAMPLES

Happening Now
I **am** ready.
She **is** ready.
They **are** ready.

Happened in the Past
I **was** ready.
Maria **was** ready.
They **were** ready.

Yesterday's Walk

Directions Circle the correct verb in each sentence that tells about the past.

1. Yesterday, Emily and Keisha (are, were) walking home from school.

2. It (was, is) raining on the walk home.

3. As they walked, they (are, were) getting wet.

4. Emily (am, was) very wet.

5. They (were, was) almost home.

6. Keisha's mom (are, was) waiting with towels when they got home.

Rhyming Words

Directions Draw lines from the words on the left to match the rhyming words on the right.

1. are   cane
2. from   mother
3. rain   car
4. other   come

A Trip to the Lake

Directions Read the letter. Use words from the box below to fill in the blanks.

| end   are   boat   about   sky   rain   lake |

Dear Carolyn,

We **(5)** _____ having fun at the

(6) _____. There is a **(7)** _____

on the water. We rode in it from one **(8)** _____

of the lake to the other. Then the **(9)** _____

got dark and cloudy. It looked like **(10)** _____,

so we went home. Our trip is just **(11)** _____

over. I will see you soon.

Your friend,

Nick

Rounding Numbers

REMEMBER

A **rounded number** is close to an exact amount. It usually ends in zero. Look at the number you are rounding to. If the number to the right of that number ends in 4 or less, round down. If that number ends in 5 or higher, round up.

EXAMPLES

Round to the nearest ten	Rounded Number	Reason
3<u>7</u>	40	7 is greater than 5, so round up.
1<u>2</u>1	120	2 is less than 4, so round down.
5<u>8</u>8	590	8 is more than 5, so round up.

Round Up or Down

Directions Round the underlined numbers to the nearest tens. Write your answers on the lines.

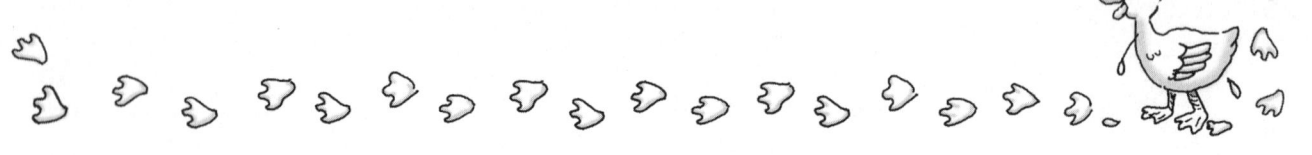

1. <u>1</u>2 _____

2. <u>1</u>7 _____

3. <u>2</u>9 _____

4. <u>4</u>5 _____

Directions Round each number. Circle **10** or **100** to show how you rounded.

5. 31 _____ feet   10 or 100

6. 189 _____ days   10 or 100

48 Math Grade 2

Time for Rain!

Directions Read the problems. Draw hands on the clocks. Then write the time on the lines.

8:00 _____

1. Paul's soccer game begins at noon. What time is it?

_____

2. He plays for 30 minutes. Then it begins to rain. The referee stops the game. What time is it?

_____

3. The rain stops in 30 minutes. The referee starts the game start again. What time is it?

_____

4. Paul's game is over after 1 more hour of play. What time is it?

_____

Math Grade 2 49

Rainy Day Puzzle

Directions Read each definition. Use words from the box below to finish the puzzle.

```
red      rain     thunder    umbrella
cloud    colors   rainbow
```

ACROSS

1. A loud noise made when it rains
4. Something colorful made when light passes through water
5. Blue, orange, and green

DOWN

2. Something used to keep the rain off
3. Water that falls from the sky
4. Color of an apple
5. Something that floats in the sky

All Wet

Directions Write sentences telling about a time that you were outside in the rain. Where were you? Did you get very wet? How did you feel? Draw a picture to go with your sentences.

Drawing/Writing Grade 2 51

CHAPTER 7
From Engine to Caboose

CHUG-chug-chug-chug, CHUG-chug-chug-chug. That's the sound a train makes when it starts moving. A train has many cars, one right after the other. The cars look like long boxes. The cars have wheels that ride on two tracks. The tracks run side by side.

The first train car is the engine. The engine pulls the train. It makes the train go fast. At the end of the train is the caboose. Between the engine and the caboose are many other cars. Some of these cars are for people to sit in. Other cars carry boxes, tools, or other things.

When you ride in a train, you get to see many things. You see towns, farms, trees, and lakes. As the train moves, you move with it. When the train chug-chugs up the side of a hill, you chug-chug with it. When the train crosses a river, you cross the river, too. Riding on a train is fun!

From Engine to Caboose

Directions Use what you have just read to answer the questions.

1. What part of the train makes it move?
   _____

2. What does the author say the shape of the train cars look like?
   _____

3. Which part of the train is last?
   _____
   _____

4. Which sentence from the article is an opinion? Circle it.
   The first train car is the engine.
   Riding on a train is fun!

5. Would you like to ride on a train? Why or why not?
   _____
   _____

Reading Comprehension *Grade 2*

Nouns and Pronouns

REMEMBER
A **pronoun** takes the place of a noun. The pronouns **I**, **you**, **he**, **she**, **it**, **we**, and **they** can take the place of nouns in the naming part of a sentence. The pronouns **me**, **you**, **him**, **her**, **it**, **us**, and **them** can take the place of nouns that follow action verbs.

EXAMPLES

Nouns

Mike wants to ride on a train.

Quan saw **airplanes** land.

Pronouns

He wants to ride on a train.

Quan saw **them** land.

Getting Packed

Directions Read each sentence. Choose a pronoun from the box below to take the place of the underlined word or words. Write the pronoun on the line.

| them | They | It | He |

1. <u>Alex and Luke</u> are going on a trip with Uncle Pete. _____

2. <u>Luke</u> can't wait to go. _____

3. The boys are packing <u>the suitcases</u>. _____

4. <u>One suitcase</u> is too full to close. _____

54 Language Grade 2

Opposite Ways

Directions Read each word. Find a word in the box below that is opposite in meaning. Write it on the line.

bus	down	give	inside
say	yellow	yours	zoo

1. outside _____
2. up _____
3. take _____
4. listen _____
5. mine _____

Where's the Zoo?

Directions Help your friend get to the zoo. Use words from the box above to finish the directions.

Step inside the (6) _____ bus. Then

(7) _____ the (8) _____

driver one dollar. Sit (9) _____ in an empty

seat. Wait for the driver to (10) _____,

"This stop is for the (11) _____." Then, walk

(12) _____ the steps carefully.

Vocabulary Grade 2

Rounding Money

REMEMBER
Money amounts are often rounded to the nearest 10¢ or to the nearest $1.00.

Round to the nearest 10¢
Round up if you have 5 or more in the ones place. Round down if you have 4 or less in the ones place.

Round to the nearest $1.00
Round up if you have 5 or more in the tens place. Round down if you have 4 or less in the tens place.

EXAMPLES

Number	Round Number	Reason
47¢	50¢	7¢ is greater than 5¢. Round up.
$2.25	$2.00	25¢ is less than 50¢. Round down.

Rounding Makes "Cents"

Directions Round each price to the nearest 10¢ or $1.00.

1.  36¢

2.  82¢

3.  $1.11

4.  $2.27

5.  $6.61

6.  $9.35

56 Math Grade 2

Packing Problems

Directions Robin is packing to go on a trip. She can't decide what to pack. She thinks about how many outfits she can make if she packs her blue sweater, her striped sweater, her sweater with a dog on it, a pair of blue jeans, and a pair of black pants. Answer the questions to help Robin pack.

1. If Robin takes all 3 sweaters and the jeans, how many outfits can she make? List them.

   _____

   _____

2. If Robin takes the blue sweater, the striped sweater, and both pairs of pants, how many outfits can she make? List them.

   _____

   _____

3. If Robin takes all 3 sweaters and both pairs of pants, how many outfits can she make? List them.

   _____

   _____

   _____

Math Grade 2

Travel Codes

Directions Look at the code. Each number stands for a letter. Write the letters on the blanks.

A	B	C	D	E	F	G	H	I	J
1	2	3	4	5	6	7	8	9	10

K	L	M	N	O	P	Q	R	S	T
11	12	13	14	15	16	17	18	19	20

U	V	W	X	Y	Z
21	22	23	24	25	26

1. What did the car say when it got new tires?

__ __ __ __ __ __ __ __ __ __
9   6   5   5   12   23   8   5   5   12

__ __ __ __.
7   15   15   4

2. Why did the train sit on a clock?

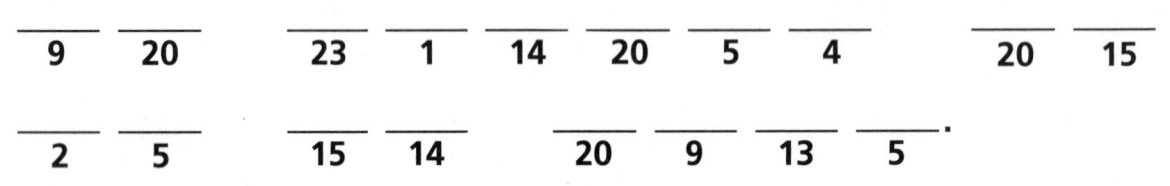

__ __ __ __ __ __ __ __ __ __
9   20   23   1   14   20   5   4   20   15

__ __ __ __ __ __ __ __.
2   5   15   14   20   9   13   5

3. Why was the picture of a train in a magazine about sports?

__ __ __ __ __ __
9   20   23   1   19   1

__ __ __ __ __ __ __ __ __.
20   18   1   3   11   19   20   1   18

4. How is a car like an elephant?

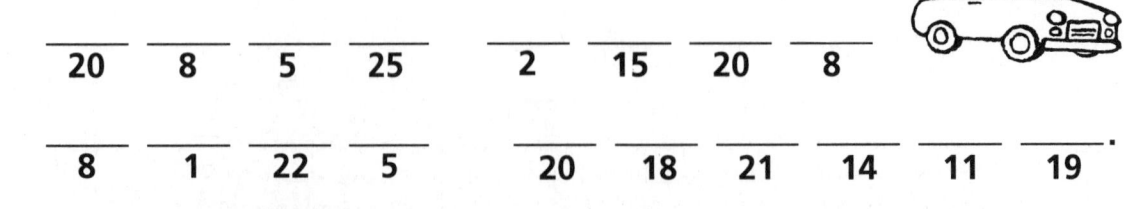

__ __ __ __ __ __ __ __
20   8   5   25   2   15   20   8

__ __ __ __ __ __ __ __ __ __.
8   1   22   5   20   18   21   14   11   19

Code Fun Grade 2

On the Move

Directions Imagine you could make a new vehicle to travel cross-country. What would it look like? Would it ride on the ground, fly in the air, or move on a track like a train? Draw a picture of your vehicle.

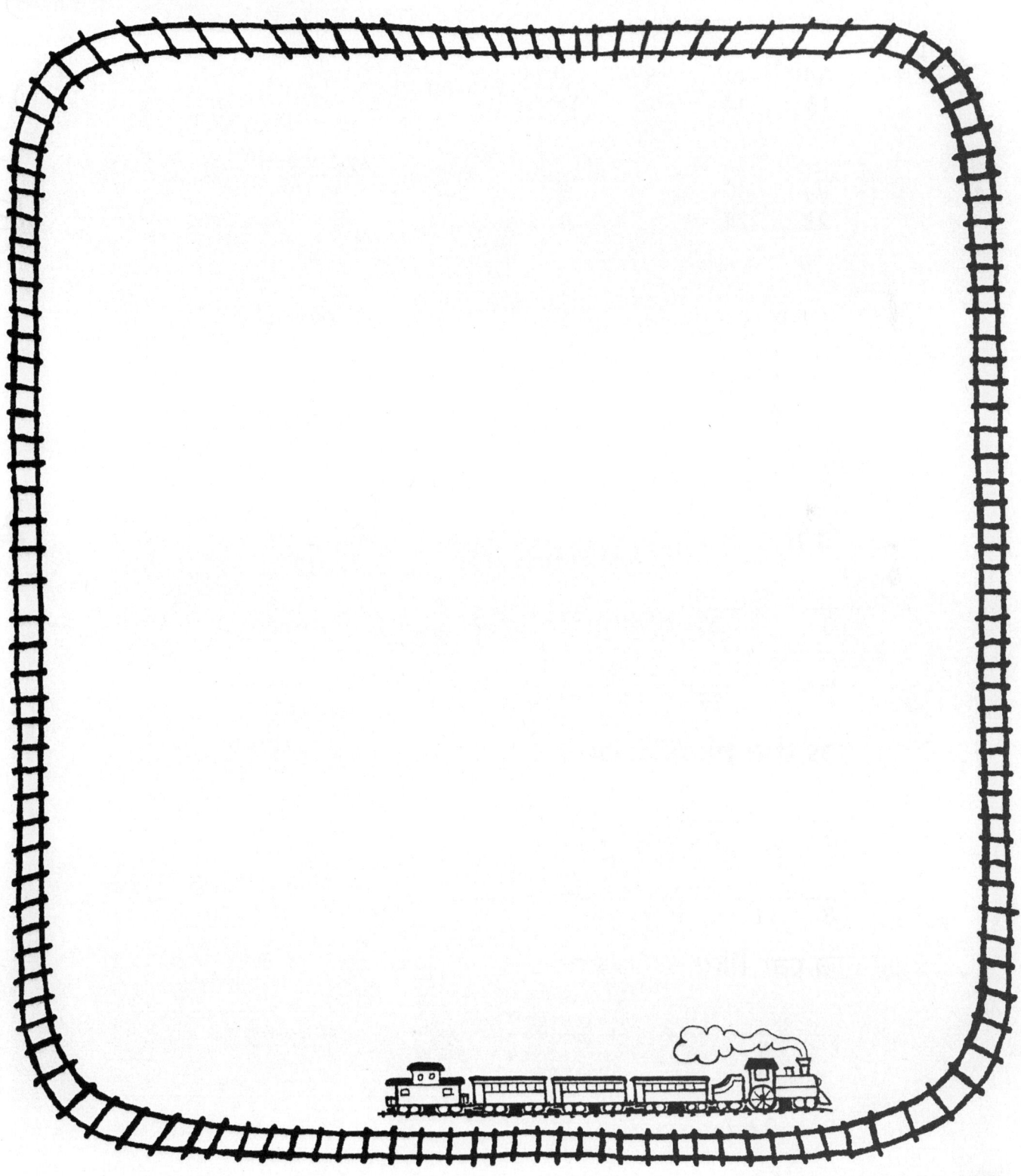

Drawing *Grade 2*

CHAPTER 8
Stars Fill the Sky

Go outside on a clear, dark night. Look up into the sky. What will you see? Stars fill the sky. They might look the same, but they are all different from each other. Some stars shine brighter than others. Their light can be different colors. In all the sky, there is one star you cannot see at night. That is the sun.

Like other stars, it is far away from the earth. The sun is closer to Earth than any other star. The sun does something the other stars do not. The sun gives us light and heat during the day. When it is night on one part of the earth, the sun gives light and heat to another part.

Stars Fill the Sky

Directions Use what you have just read to answer the questions.

1. What is this article mainly about?
   _____

2. What star cannot be seen at night?
   _____

3. Where does this star go?
   _____

4. Name one way the sun is like other stars.
   _____

5. Name one way the sun is different from other stars.
   _____
   _____

Reading Comprehension Grade 2

Adjectives

REMEMBER

An **adjective** is a word that tells about a noun. It may tell how something looks, tastes, smells, sounds, or feels.

EXAMPLES

The moon is **bright**.
The sun looks **yellow**.

What Is It Like?

Directions Circle the adjective in each sentence.

1. The sky was blue.

2. The clouds are fluffy.

3. The sun was hot.

4. The stars look small.

5. We saw a big cloud.

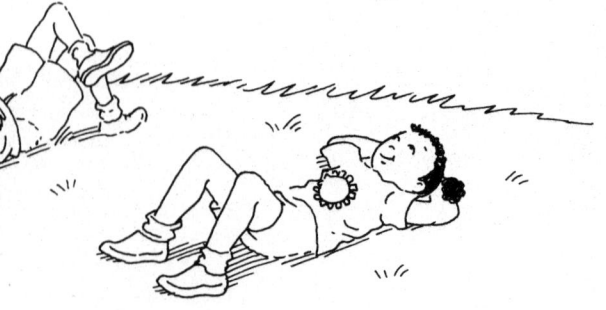

Rhyming Words

Directions Draw lines from the words on the left to their rhyming words on the right.

1. black   night
2. bee   star
3. far   soon
4. light   day
5. way   track
6. moon   see

Night Flight

Directions Read the poem. Use words from the box to fill in the blanks.

| moon | away | light | could | black |

A **(7)** _____ and yellow bee one day,

went buzz, buzz, buzzing far **(8)** _____.

At night time when the sun went down,

a friendly hive **(9)** _____ not be found.

The **(10)** _____ felt sorry for the bee,

and shined her **(11)** _____ for it to see.

Vocabulary Grade 2

Make an Estimate

REMEMBER

Some groups of things can be counted. With other groups, you must take a guess at the amount because there are too many to count. An **estimate** is a guess at the number of things in a large group.

EXAMPLES

You would **count** the number of chairs in the classroom. You would **estimate** the number of clouds in the sky.

Count or Guess?

Directions Write **count** if the items can be counted. Write **estimate** if they must be estimated.

1. hours in a day _____
2. hairs on your head _____
3. number of fish in a bowl _____
4. days until Saturday _____
5. jelly beans in the jar _____

Directions Complete each sentence with an estimate.

6. I drink about _____ glasses of water each day.
7. I read about _____ books last year.
8. My heart beats about _____ times each minute.

Spinning in Space

Directions At science camp, the campers were learning about space travel. Mr. Hile is a counselor at the camp. He asked the campers to help make a mural of space. Each camper got to spin the arrow to find out what part of the mural they would draw. Use the spinner to answer the questions.

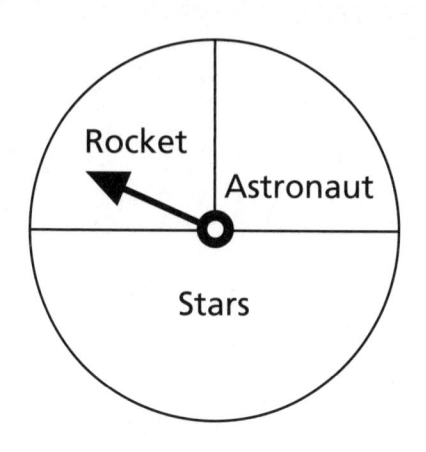

1. How many parts does the spinner have? _____

2. Which part is the largest? _____

3. Marvin spins the arrow. On which word is the arrow most likely going to stop? Tell why you think so. _____
   _____

4. Twelve campers spin the arrow. About how many campers will make a star? Tell why you think so. _____
   _____

5. Out of the twelve campers, how many will make a rocket?
   _____

Math Grade 2

Space Word Search

Directions Find the words listed in the box below. Then circle them in the puzzle. The words are hidden across and down.

stars	moon	rocket	planet
sun	earth	cloud	space

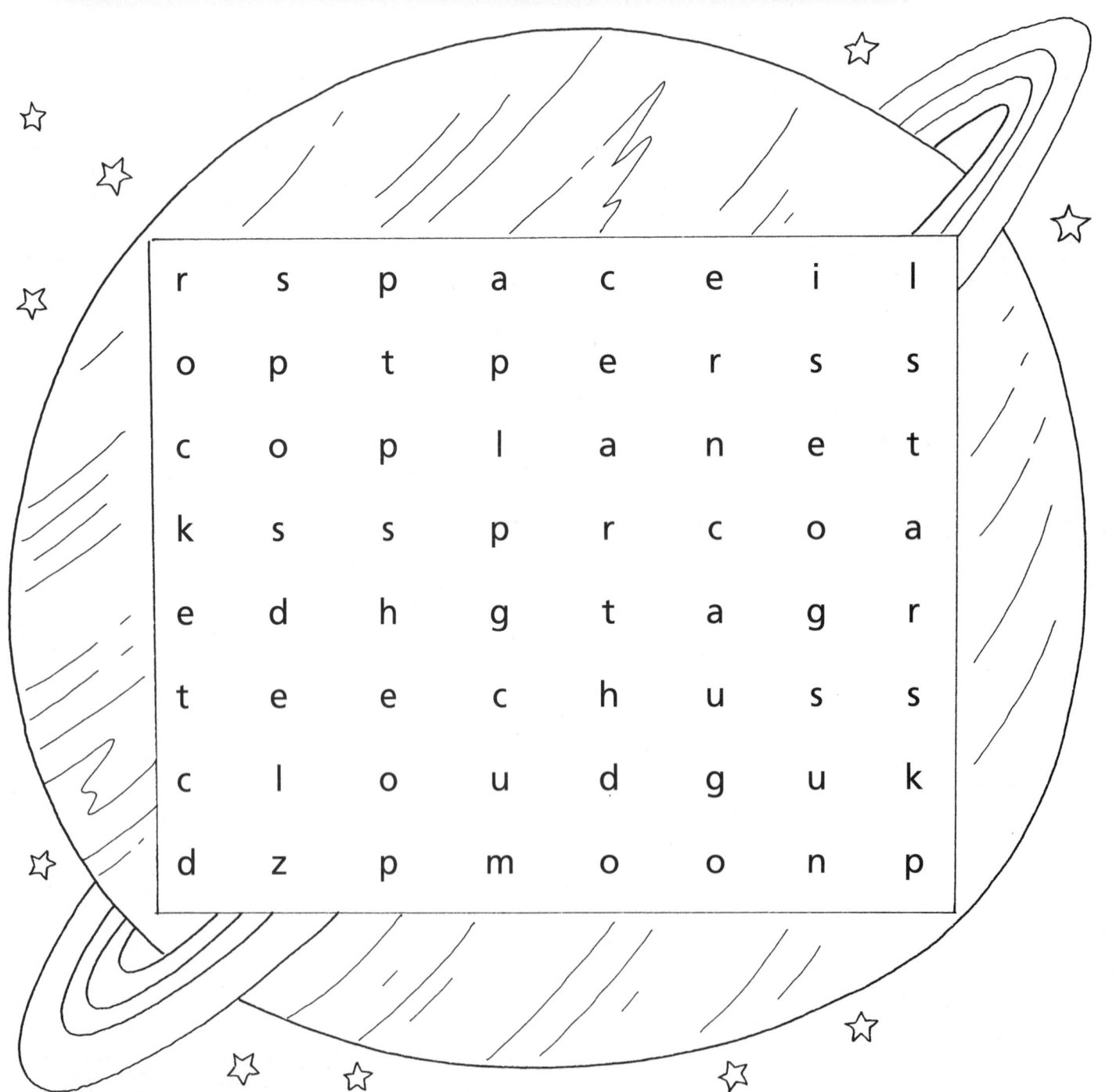

r	s	p	a	c	e	i	l
o	p	t	p	e	r	s	s
c	o	p	l	a	n	e	t
k	s	s	p	r	c	o	a
e	d	h	g	t	a	g	r
t	e	e	c	h	u	s	s
c	l	o	u	d	g	u	k
d	z	p	m	o	o	n	p

66 Word Search *Grade 2*

Creature Features

Directions Suppose you rode a rocket into space. What kind of space creature might you meet? What would it look like? Draw a picture of the space creature. Then write sentences telling about it.

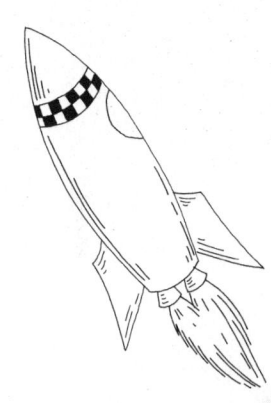

_____

_____

_____

_____

Drawing/Writing Grade 2

CHAPTER 9

Spider Tricks Snake

Long ago, Spider had a wish. He wanted to catch Snake. He had tried before, but never caught him. Spider went to see Tiger, the animal king, to ask him how to catch Snake. Spider told Tiger what he wanted. Tiger told Spider that Snake was very clever. Spider was going to have to trick Snake to catch him.

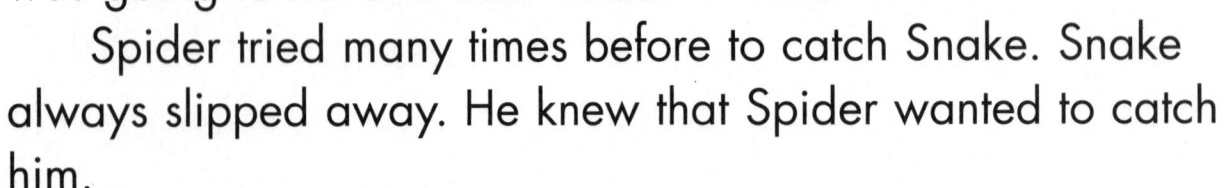

Spider tried many times before to catch Snake. Snake always slipped away. He knew that Spider wanted to catch him.

One day Spider said to Snake, "I want to see if you are longer than a bamboo pole."

This hurt Snake's feelings. "Of course I am!" Snake said.

"Then lie down by the pole," said Spider. "But you can't move up or down while I look," said Spider.

"Then tie me to the pole," said Snake. So Spider tied him to the pole. And that was Spider's trick. That's how he caught Snake.

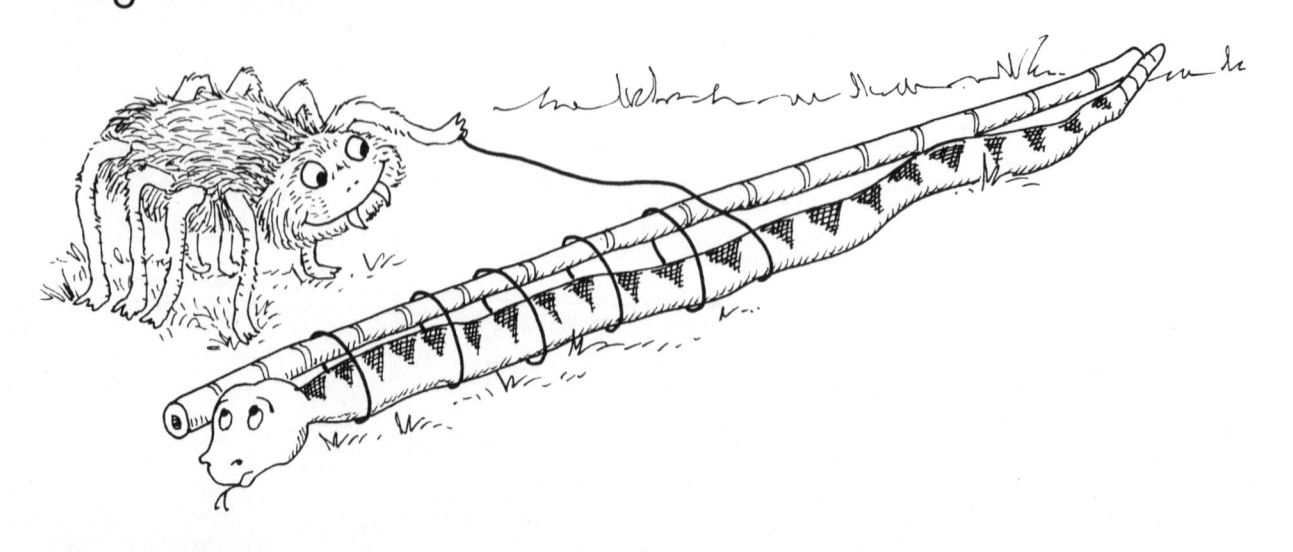

Spider Tricks Snake

Directions Use what you have just read to answer the questions.

1. What did Spider want?

   _____

2. How did Tiger help Spider catch Snake?

   _____

3. How did Spider hurt Snake's feelings?

   _____

4. What might Snake do the next time Spider asks him to do something?

   _____

5. Is this selection real or make-believe? How do you know?

   _____

   _____

Reading Comprehension *Grade 2*

Telling Sentences

REMEMBER
A **telling sentence** tells something. The sentence begins with a capital letter. It ends with a period.

EXAMPLES
My horse can do tricks.
He walks on his back legs.

Tell All About It

Directions Draw a line under the correct telling sentence.

1. Mark has a dog
   Mark has a dog.

2. his dog's name is Tiger.
   His dog's name is Tiger.

3. Tiger can do tricks.
   tiger can do tricks

4. mark throws a small ball.
   Mark throws a small ball.

5. Tiger jumps into the air and catches the ball.
   Tiger jumps into the air and catches the ball

70 Language *Grade 2*

The Same Game

Directions Read each word. Find a word in the box below that has the same or almost the same meaning and write it on the line.

| trick | funny | grade | hear |
| job | house | think | story |

1. work _____
2. listen _____
3. joke _____
4. tale _____
5. silly _____
6. home _____

Rabbit Tales

Directions Read the story. Use words from the box above to fill in the blanks.

I am in second **(7)** _____. I like to read books that are **(8)** _____. When I go to the library, I always look for a **(9)** _____ that will make me laugh. I found some books that are Native American tales. The animals in these stories like to **(10)** _____ each other. I **(11)** _____ the stories that have Rabbit in them are best. Poor Rabbit always gets into trouble.

Vocabulary Grade 2   71

Adding: Carry the One

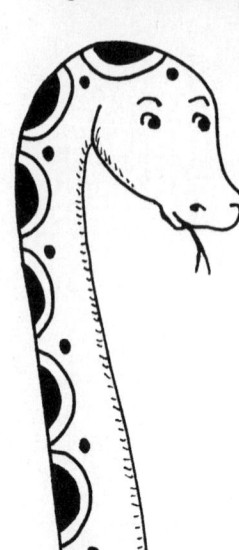

REMEMBER

When you **add**, you put groups of things together. To show addition, you write **+**. This sign is a **plus sign** or an **addition sign**. The answer is called the **sum** or **total**. Sometimes, you will need to **regroup 1** or **carry 1** to the tens place.

EXAMPLES

```
                                1 ← regrouped 1
    5          15              26
  +4          +4              +8
  ---         ---             ---
    9          19              34
```

Add It Up

Directions Find each sum. Regroup if you need to.

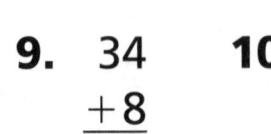

1. 3   2. 4   3. 9   4. 8
   +1   +4   +3   +6

5. 14   6. 17   7. 14   8. 17
   +5   +2   +6   +8

9. 34   10. 46   11. 29   12. 57
   +8   +5   +6   +7

Twenty-Pound Package

Directions Emilio is going to mail the gifts below, but he can only use two boxes. Each box can hold no more than 20 pounds. Answer the questions to help Emilio decide the best way to mail the gifts.

books	roller skates	radio	tools
6 pounds	7 pounds	2 pounds	17 pounds

1. What is the total weight of the books and the roller skates? _____

2. What is the total weight of the radio and the roller skates? _____

3. If Emilio puts 2 gifts in each box, which gifts can be mailed together? Name two.

   _____ and _____

   _____ and _____

4. Emilio finds a larger box that can hold up to 40 pounds. Can Emilio mail all 4 gifts in this box? Tell why or why not.

   _____

   _____

Math Grade 2

Tricky Riddles

Directions Use the clues to solve the riddles. Hint: Think about words that rhyme with the words in dark print.

EXAMPLE

What kind of a **tool** is a pencil?

s c h o o l   t o o l

1. What do you call a **hat** that a magician sat on?

   ___ ___ ___   h a t

2. What do you call the wand that a magician uses to help do a **trick**?

   t r i c k   ___ ___ ___ ___ ___

3. What would you call a magician's **cape** if it were purple?

   ___ ___ ___ ___   c a p e

4. What do you call a noisy **crowd**?

   ___ ___ ___ ___   c r o w d

Riddles *Grade 2*

What a Trick!

Directions Have you ever seen a magician perform a trick? Do you know how it was done? Take a guess! Write sentences telling about the trick and how you think the magician did the trick.

CHAPTER 10

The Gingerbread Man

One day, a little old lady made a gingerbread man. He would be good to eat. She went to pop the gingerbread man into the oven. Just then, he jumped off the pan. He ran out of the house. The lady chased him. He said, "Run, run as fast as you can. You can't catch me. I'm the gingerbread man!"

A cow chased the gingerbread man. So did a horse. A dog joined in. Then the gingerbread man got to a river. How could he cross? He saw a fox.

"Quick!" said the Fox. "I will help you cross the river. You can ride on my back."

The gingerbread man was glad to have the fox's help. He didn't know the fox wanted to eat him, too. He jumped on the fox's back. They went into the river. The water got deep. The fox said to move up to his nose. When the gingerbread man moved, the fox snapped him up. He was good to eat!

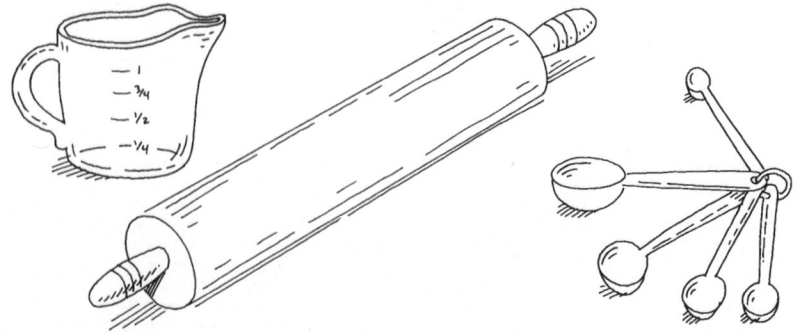

The Gingerbread Man

Directions Use what you have just read to answer the questions.

1. Why did the little old lady make the gingerbread man?
   _____

2. Who chased the gingerbread man? Name two.
   _____

3. Why did the gingerbread man get on the fox's back?
   _____
   _____

4. What clue tells you that the fox wanted to eat the gingerbread man?
   _____
   _____

5. What else could the gingerbread man have done to cross the river?
   _____
   _____

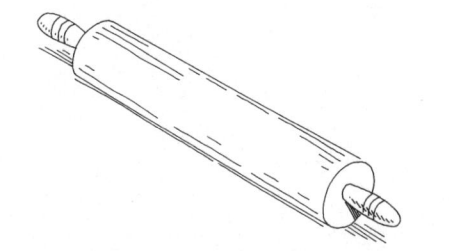

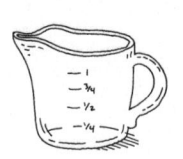

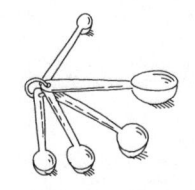

Ask a Question

REMEMBER

A **question** asks something. A question begins with a capital letter. It also ends with a question mark.

EXAMPLES

May I help you?
How long will the muffins cook?

Cooking Up Questions

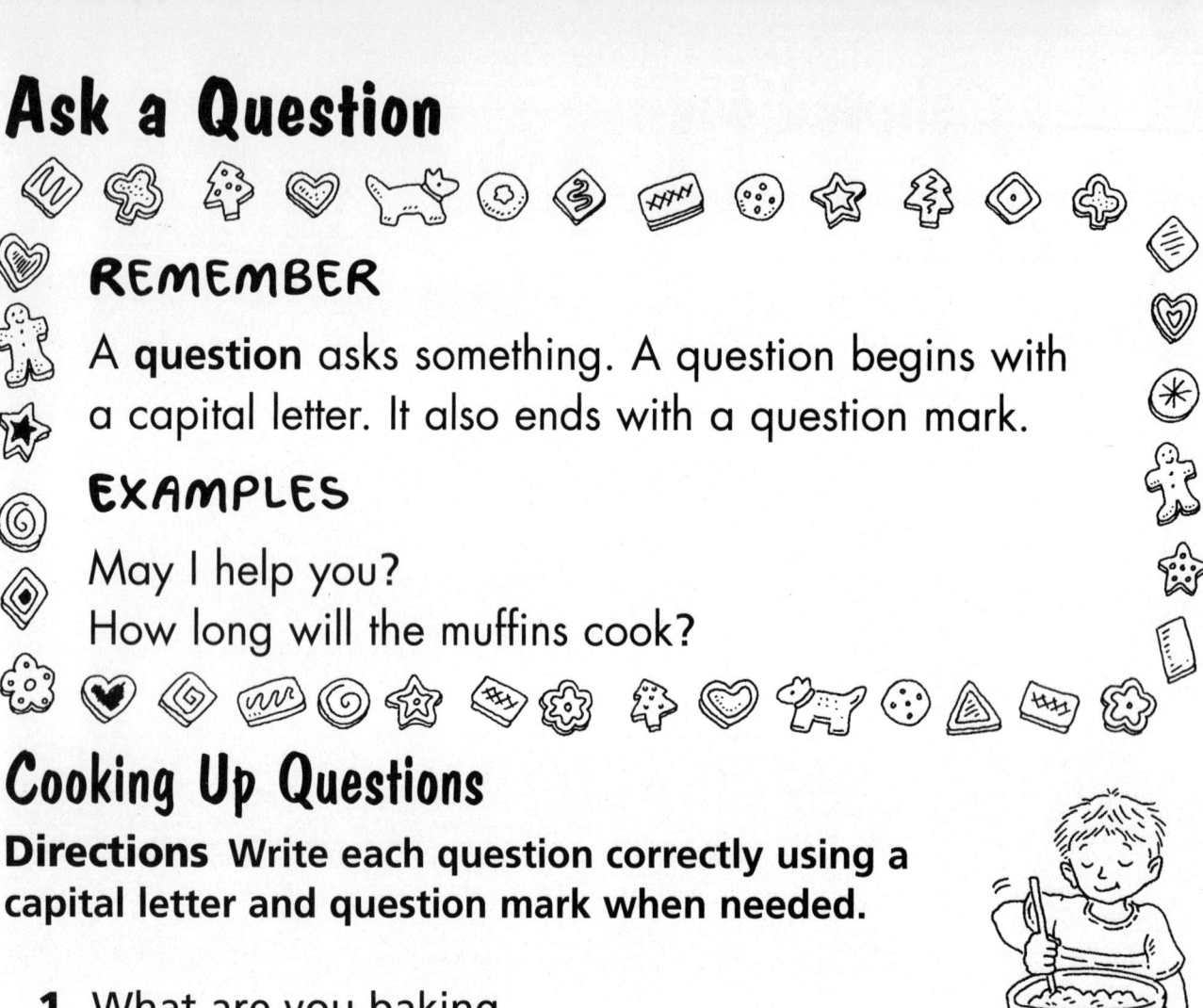

Directions Write each question correctly using a capital letter and question mark when needed.

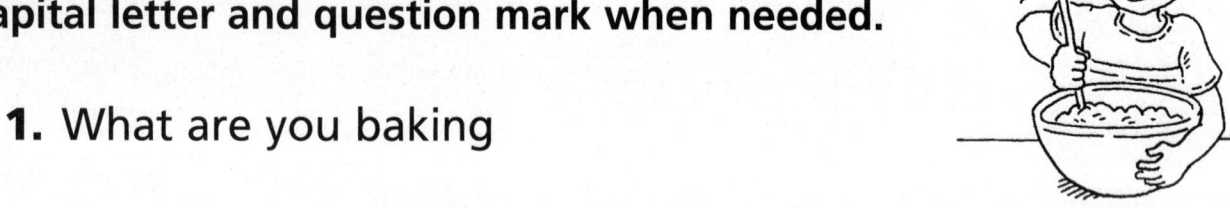

1. What are you baking
   _____

2. can I lick the spoon?
   _____

3. What time will we eat
   _____

4. may I set the table
   _____

5. Would you pass the peas, please
   _____

Kitchen Connection

Directions Draw lines from the words on the left to match their clues on the right.

1. wash — to cook in an oven
2. add — to clean with water
3. bake — one more than three
4. four — to put together

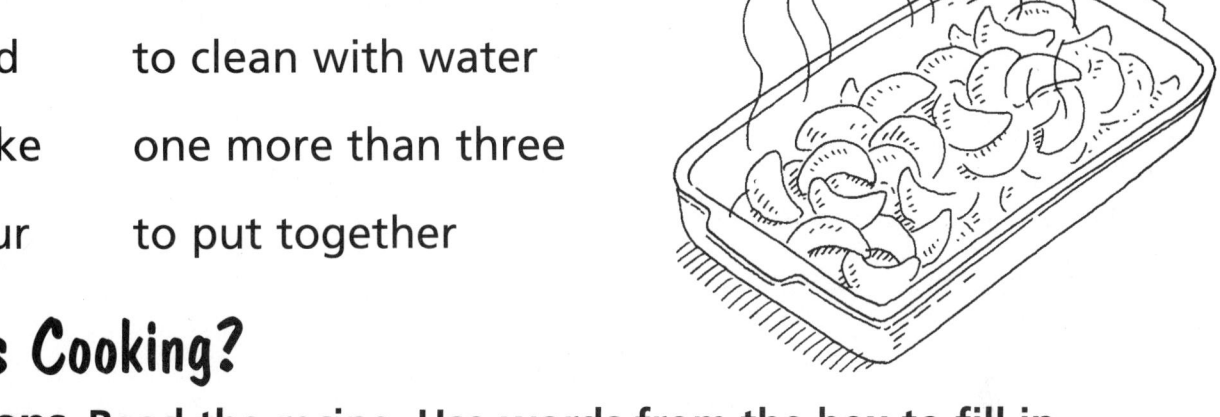

What's Cooking?

Directions Read the recipe. Use words from the box to fill in the blanks.

| cup | Add | Bake | cut | Four | need |

Baked Apple Recipe

You will **(5)** _____:

    (6) _____ apples   1 cup oatmeal

    1/2 **(7)** _____ brown sugar   3/4 cup hot water

1. Wash the apples.
2. Have an adult **(8)** _____ the apples into small pieces.
3. Mix the sugar and hot water.
4. **(9)** _____ a little butter to the mixture.
5. Pour the mixture over the apples
6. **(10)** _____ for one hour. Yum!

Vocabulary Grade 2

Subtract: Find the Difference

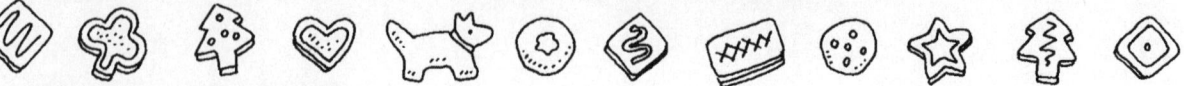

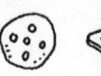

REMEMBER

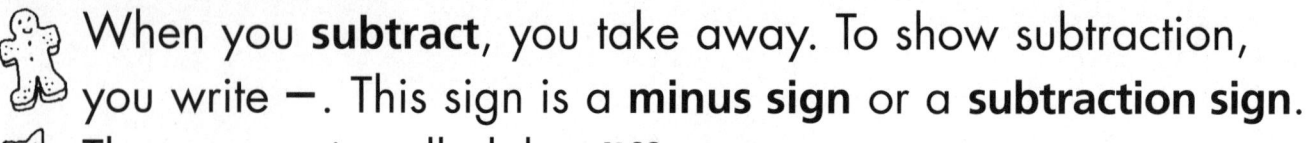

When you **subtract**, you take away. To show subtraction, you write **−**. This sign is a **minus sign** or a **subtraction sign**. The answer is called the **difference**.

Sometimes, you will need to **regroup 10** to the ones place.

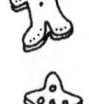

EXAMPLES

```
                                    2 11  ← regrouped 10
    9           17                   3̷1̷
  − 4         −  4                 − 14
  ───         ────                 ────
    5           13                   17
```

Take It Away

Directions Find each difference. Regroup if you need to.

1. 4   **2.** 7   **3.** 8   **4.** 9
      − 1   − 3   − 5   − 0

5. 14   **6.** 14   **7.** 17
      − 5   − 6   − 5

8. 21   **9.** 24   **10.** 40
      − 4   − 8   − 19

80 Math Grade 2

Lines of Symmetry

REMEMBER

A **line of symmetry** divides a shape into two parts. If you fold the shape along the line of symmetry, each part matches exactly.

EXAMPLE

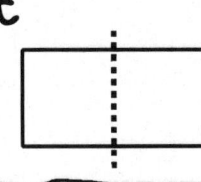

Parts and Pieces

Directions Circle **yes** if the dotted lines show symmetry. Circle **no** if they do not.

1. yes or no

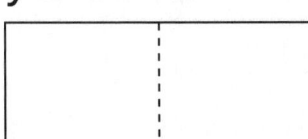

2. yes or no

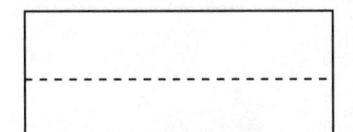

3. yes or no

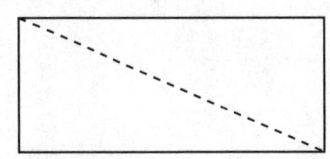

Directions Draw a line of symmetry for each shape.

4.

5.

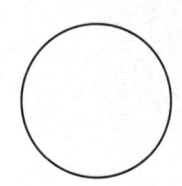

6.

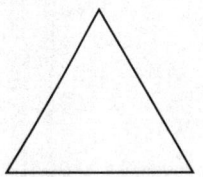

Directions Draw a line that is **not** a line of symmetry for each shape.

7.

8.

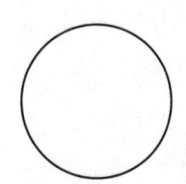

9.

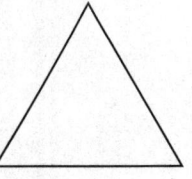

Math Grade 2   81

Food Scramble

Directions Read each meaning. Then unscramble the letters to find the words. Write the words on the lines.

1. papel   A fruit that can be red, green, or yellow

   __ __ __ __ __
       1

2. eokcoi   A baked treat that is small and flat

   __ __ __ __ __ __
      2

3. troacr   A long, orange vegetable

   __ __ __ __ __ __
    3   4

4. nove   Something food is baked in

   __ __ __ __
       5

Directions Answer the riddle. Write the letters on the lines to match the numbers above.

What did the baby corn call his dad?

__ __ __   __ __ __ __ __
1   2   1   3   2   4   5

82   Word Scramble *Grade 2*

Favorite Food Fun

Directions What is your favorite food? Draw a picture of it, but don't let anyone see it. Then write four clues about the food. Read your clue to someone. Ask the person to guess your favorite food. Show them your picture.

READING CHECK-UP
Brave Firefighters

Do you smell smoke? Is that a fire truck rushing by? If there is a fire, firefighters put it out. Fires burn quickly. That's why firefighters must work quickly. They use ladders and hoses to put out fires. They help get people out of burning buildings. They try to save homes and other places.

Firefighters wear big coats, hats, and boots. These clothes help them stay safe. Even with the clothes, firefighters might get hurt. They still do their job though. Firefighters are brave!

Directions Fill in the bubble next to the correct answer.

1. Which of these is a clue that there is a fire?
   - (A) You have to work quickly.
   - (B) You see the sun.
   - (C) You wear a hat, coat, and boots.
   - (D) You smell smoke.

2. Which of these would a firefighter **not** use on the job?
   - (A) truck
   - (B) ladder
   - (C) paintbrush
   - (D) hose

3. Firefighters wear a hat, coat, and boots to
   - (A) look good.
   - (B) keep safe.
   - (C) keep warm.
   - (D) go faster.

4. In this selection, **rushing** means
   - (A) going fast.
   - (B) being loud.
   - (C) working hard.
   - (D) calling for help.

READING CHECK-UP
Check It Out!

Where would you go to check out books? You would go to the library. You might be able to check out other things, too. In some libraries, you can check out music or videotapes. Some even have paintings to loan. Just like books, you have to return them too. Most libraries want the items back in two to three weeks. You must take good care of the things you borrow. Other people will want to borrow them when you are done.

Directions Fill in the bubble next to the correct answer.

5. What is another good title for this article?

- Ⓐ At the Library
- Ⓑ Places in the Neighborhood
- Ⓒ How Books Are Made
- Ⓓ The Best Book Ever

6. In the article, **loan** means to

- Ⓐ ask someone for something.
- Ⓑ fix something.
- Ⓒ let someone have something for a while.
- Ⓓ pick out a book.

7. Which of these would you **not** find at a library?

- Ⓐ books
- Ⓑ tapes
- Ⓒ paintings
- Ⓓ pets

8. For how long do most libraries loan books?

- Ⓐ two days
- Ⓑ two to three weeks
- Ⓒ two months
- Ⓓ two years

READING CHECK-UP
Jack and the Giant

Once upon a time, there was a boy named Jack. He lived in a castle with a lazy giant. The giant ate all of Jack's food and snored very loudly.

One day Jack and the giant walked to town. They met a clown. Jack asked if the clown wanted a giant for his circus. The clown was very happy. Jack traded the giant for a bag of corn.

Jack went home, planted the corn, and went to sleep. The next day, he saw a tall cornstalk. "Not again!" shouted Jack.

Jack climbed the cornstalk. When he reached the top, he saw his cow Mabel. Jack had traded her long ago for some magic beans. Jack and Mabel climbed down the cornstalk and returned to the castle.

Directions Fill in the bubble next to the correct answer.

9. What did Jack trade for the giant?
   - Ⓐ magic beans
   - Ⓑ a bag of corn
   - Ⓒ a cow
   - Ⓓ magic sunflower seeds

10. How did the clown feel to get the giant?
    - Ⓐ angry
    - Ⓑ sick
    - Ⓒ glad
    - Ⓓ sleepy

11. Why couldn't Jack sleep at night?
    - Ⓐ The giant liked to stay up.
    - Ⓑ Mabel snored.
    - Ⓒ He was busy planting.
    - Ⓓ The giant snored.

12. The giant and the cow are alike because they both
    - Ⓐ were traded.
    - Ⓑ lost their way.
    - Ⓒ joined the circus.
    - Ⓓ snored.

READING CHECK-UP
Hurray for Hummingbirds!

Hummingbirds are the smallest birds in the world. They are very special birds, too. Hummingbirds can do two things that no other birds can do.

A hummingbird is the only bird that can fly backwards. It also can hover, or park, in the air. A hummingbird hovers in front of a flower. Then it puts its long beak into the flower for a drink.

Can you guess why this bird is called a hummingbird? It is named for the humming sound it makes. A hummingbird beats its wings so fast that the wings make a humming sound. In fact, it beats its wings so fast that you cannot see them.

Directions Fill in the bubble next to the correct answer.

13. What is another title for this selection?
- Ⓐ Big Birds
- Ⓑ Hummingbird Food
- Ⓒ Special Birds
- Ⓓ Bird Sounds

14. In the article, **hover** means to
- Ⓐ stay in one place.
- Ⓑ park a car.
- Ⓒ hum a song.
- Ⓓ fly backwards.

15. Which is **not** a fact about hummingbirds?
- Ⓐ They can fly backwards.
- Ⓑ Their wings beat very fast.
- Ⓒ They have short beaks.
- Ⓓ They are the smallest birds.

16. How did the hummingbird get its name?
- Ⓐ Its wings make a humming sound.
- Ⓑ It doesn't make a sound.
- Ⓒ It flies to humming sounds.
- Ⓓ It flies very slowly.

STOP! Number correct: _____ out of 16

MATH CHECK-UP
Directions Answer each question.

1. Which equals 7?
- Ⓐ 6 − 1
- Ⓑ 9 − 2
- Ⓒ 8 − 2
- Ⓓ 7 − 1

2. Which number is even?
- Ⓐ 1
- Ⓑ 3
- Ⓒ 4
- Ⓓ 7

3. Fill in the blank.
2, 4, ____, 8, 10
- Ⓐ 0
- Ⓑ 5
- Ⓒ 6
- Ⓓ 7

4. What does the underlined number stand for? 32̲1
- Ⓐ ones
- Ⓑ tens
- Ⓒ hundreds
- Ⓓ thousands

5. Which sign would make the number sentence true?

5+1 ☐ 4+2
- Ⓐ =
- Ⓑ −
- Ⓒ >
- Ⓓ <

6. What time does the clock show?
- Ⓐ 12:00
- Ⓑ 12:15
- Ⓒ 2:30
- Ⓓ 3:00

7. How many cents does the coin stand for?
- Ⓐ 1¢
- Ⓑ 5¢
- Ⓒ 10¢
- Ⓓ 25¢

8. Which shape is a square?
- Ⓐ
- Ⓑ
- Ⓒ
- Ⓓ

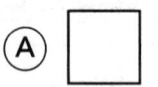

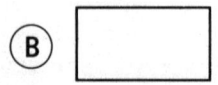

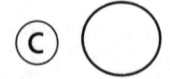

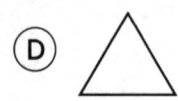

MATH CHECK-UP
Directions Answer each question.

9. 5 + 3 =
- Ⓐ 2
- Ⓑ 5
- Ⓒ 7
- Ⓓ 8

10. 7 − 4 =
- Ⓐ 3
- Ⓑ 5
- Ⓒ 11
- Ⓓ 12

11. 12 − 9 =
- Ⓐ 3
- Ⓑ 5
- Ⓒ 11
- Ⓓ 21

12. 7 + 7 =
- Ⓐ 0
- Ⓑ 7
- Ⓒ 12
- Ⓓ 14

13. 12
   + 9
- Ⓐ 3
- Ⓑ 11
- Ⓒ 21
- Ⓓ 27

14. 44
   − 5
- Ⓐ 49
- Ⓑ 39
- Ⓒ 29
- Ⓓ 19

15. 36
   + 11
- Ⓐ 57
- Ⓑ 47
- Ⓒ 35
- Ⓓ 25

16. 58
   + 24
- Ⓐ 34
- Ⓑ 64
- Ⓒ 72
- Ⓓ 82

17. 50
   − 21
- Ⓐ 71
- Ⓑ 39
- Ⓒ 31
- Ⓓ 29

Math Check-Up *Grade 2*

MATH CHECK-UP

18. Cindy began practicing the piano at 4:15. P.M. She practiced for 35 minutes. What time did she finish?

- Ⓐ 4:35
- Ⓑ 4:40
- Ⓒ 4:50
- Ⓓ 5:15

19. Isaac has a quarter and two nickels. How much money does he have?

- Ⓐ 25¢
- Ⓑ 27¢
- Ⓒ 30¢
- Ⓓ 35¢

20. Round this number to the nearest ten.

1<u>3</u>4

- Ⓐ 100
- Ⓑ 130
- Ⓒ 135
- Ⓓ 140

21. What time does the clock show?

- Ⓐ 4:00
- Ⓑ 3:15
- Ⓒ 4:15
- Ⓓ 5:00

22. Tia baked 24 muffins. Her family ate 6 of them. How many were left?

- Ⓐ 22
- Ⓑ 19
- Ⓒ 18
- Ⓓ 12

23. Which of these items has the shape of a cylinder?

- Ⓐ soup can
- Ⓑ basketball
- Ⓒ cereal box
- Ⓓ book

STOP!

Number correct: _____ out of 23

Congratulations!

_____
(name)

has completed *Summer Counts!*

Good job!

Have a good school year.

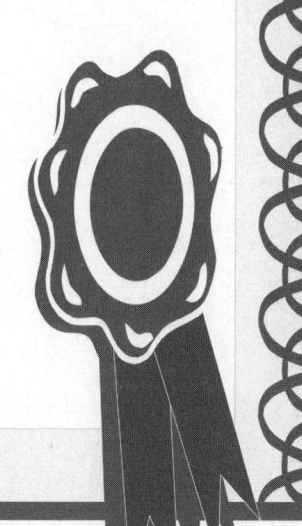

Answer Key

Page 5
1. Possible answer: The American Flag.
2. The flag has 50 stars.
3. The stars stand for the 50 states.
4. Possible answers: post office, firehouse, classroom, library, house, parades.
5. Real. It gives true facts about the American flag.

Page 6
1. Ana–person; parade–thing
2. flag–thing
3. friends–people
4. Ana–person; school–place
5. parade–thing; park–place
6. town–place

Page 7
1. school
2. class
3. flag
4. friends
5. school
6. class
7. teacher
8. flag
9. friends

Page 8
1. five
2. two
3. seven
4. four
5. ten
6. nine
7. one
8. six
9. eight
10. three
11. 3; 5; 8; 9
12. 6; 10; 14; 18; 20
13. 15; 20; 30; 45; 50
14. 20; 40; 60; 80; 90

Page 9
1. odd
2. even
3. odd
4. even
5. even
6. odd
7. even
8. odd
9. odd
10. even

Page 10
Across
2. blue
3. stripes
6. flag
7. states

Down
1. country
4. stars
5. white

Page 13
1. The story takes place in a neighborhood park.
2. Possible answers: they ate a picnic, fed the birds, tossed a ball, picked up trash, and took a picture.
3. Yes. The story says that everyone smiled and laughed.
4. He means that you should take everything you brought to the park home with you.
5. 2, 3, 1

Page 14
1. We live in Elmwood.
2. My favorite park is on Beech Street.
3. My friend Alex lives near the park.
4. We meet at the park on Saturday.

Page 15
1. outside
2. stay
3. after
4. take
5. love
6. sit
7. sky
8. cloudy
9. park
10. run
11. outside
12. side

Page 16
1. tens
2. hundreds
3. ones
4. tens
5. hundreds
6. 164
7. 315
8. 628

Page 17
1. 14
2. 146
3. 64
4. 641
5. 461
6. 14; 16; 41; 46

Page 18
1. race
2. outside
3. laugh
4. swing
Riddle: rough, rough.

Page 21
1. It is about a cat who tricks mice out of cheese.
2. They wanted to each have the same amount of cheese.
3. The story said that the cat had a plan.
4. Possible answer: The mice were mad because they were tricked.
5. Make-believe. Cats and mice cannot really talk.

Page 22
1. live
2. opens
3. swings
4. glow
5. sing

Page 23
1. sheep
2. farm
3. dear
4. saw
5. farm
6. many
7. saw
8. very
9. goat
10. milk

Page 24
1. 67¢
2. $1.34
3. 49¢
4. $1.24

Page 25
1. 50¢
2. $1.50
3. one dollar and eighty-five cents; $1.85
4. six quarters or five quarters, two dimes, and one nickel.
5. thirty-five cents; 35¢
6. banana

Page 26

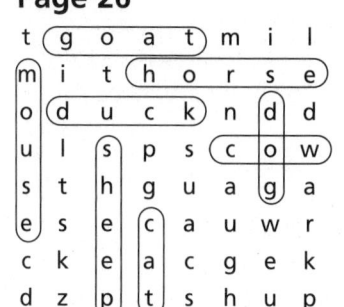

```
t g o a t m i l
m i t h o r s e
o d u c k n d d
u l s p s c o w
s t h g u a g a
e s e c a u w r
c k e a c g e k
d z p s h u p
```

Answer Key Grade 2

Page 29
1. Order: 4, 1, 2, 3.
2. If the steps are not in order, the seeds might not grow.
3. You need to give them water and sun.
4. A sprout is a small plant that is just beginning to grow from the seed.
5. Drawings will vary. Accept reasonable drawings.

Page 30
1. asks
2. wants
3. plant
4. water
5. looks
6. grow

Page 31
1. any
2. tall
3. our
4. grow
5. tree
6. than
7. plant
8. seed
9. any
10. plant
11. grow
12. tree
13. tall
14. than
15. our

Page 32
1. <
2. >
3. <
4. <
5. Address order starting at top left box: 128, 174, 200, 242, 278.

Page 33
1. 70
2. 50
3. a. The thermometer shows 40°F.
3. b. The thermometer shows 75°F.
3. c. The thermometer shows 60°F.

Page 34
1. Leaf me alone.
2. It was berry happy.
3. By its bark.
4. I'm branching out.

Page 37
1. A neighborhood is where people live, work, and play.
2. There are busy neighborhoods and quiet neighborhoods.
3. Answers will vary.
4. Possible answer: A farm can be different because one family may live on a farm, and many families live in a neighborhood.
5. People live in a neighborhood.

Page 38
1. visited
2. walked
3. learned
4. talked
5. enjoyed
6. asked

Page 39
1. it has two wheels
2. to make something work right
3. tripped over
4. not on
5. riding
6. bike
7. fell
8. chain
9. fix
10. thank

Page 40
1. <
2. >
3. <
4. <
5. <
6. >
7. 49¢
8. $1.28

Page 41
1. circle graph
2. 4 kinds of food
3. lemonade
4. apples
5. $2.25

Page 45
1. Order: 3, 2, 1.
2. People mostly see rainbows after it rains.
3. Rainbows are made when light shines through drops of water in the air.
4. Possible answers: red, orange, green, yellow, blue, purple.
5. Real. It tells facts about rainbows.

Page 46
1. were
2. was
3. were
4. was
5. were
6. was

Page 47
1. car
2. come
3. cane
4. mother
5. are
6. lake
7. boat
8. end
9. sky
10. rain
11. about

Page 48
1. 10
2. 20
3. 30
4. 50
5. 30-rounded to 10
6. 190-rounded to 10 or 200-rounded to 100

Page 49
1. 12:00
2. 12:30
3. 1:00
4. 2:00

Page 50
Across
1. thunder
4. rainbow
5. colors
Down
2. umbrella
3. rain
4. red
5. cloud

Page 53
1. The engine makes the train move.
2. The train cars look like boxes.
3. The caboose is the last part of the train.
4. Riding on a train is fun!
5. Answers will vary. Accept reasonable responses.

Page 54
1. They
2. He
3. them
4. It

Page 55
1. inside
2. down
3. give
4. say
5. yours
6. yellow
7. give
8. bus
9. down
10. say
11. zoo
12. down

Page 56
1. 40¢
2. 80¢
3. $1.10 or $1.00
4. $2.30 or $2.00
5. $6.70 or $7.00
6. $9.40 or $9.00

Page 57
1. 3 outfits: blue sweater and jeans; striped sweater and jeans; dog sweater and jeans
2. 4 outfits: blue sweater and jeans; striped sweater and jeans; blue sweater and black pants; striped sweater and black pants
3. 6 outfits: blue sweater and jeans; striped sweater and jeans; dog sweater and jeans; blue sweater and black pants; striped sweater and black pants; dog sweater and black pants

Page 58
1. I feel wheel good.
2. It wanted to be on time.
3. It was a track star.
4. They both have trunks.

Page 61
1. It tells about the sun.
2. The sun cannot be seen at night.
3. It is on the other side of the earth.
4. Possible answer: It is far away from Earth.
5. Possible answer: The sun gives Earth light and heat.

Page 62
1. blue
2. fluffy
3. hot
4. small
5. big

Page 63
1. track
2. see
3. star
4. night
5. day
6. soon
7. black
8. away
9. could
10. moon
11. light

Page 64
1. count
2. estimate
3. count
4. count
5. estimate
6–8. Answers will vary. Accept reasonable responses.

Page 65
1. 3
2. stars
3. It will most likely stop on stars because it is the largest part.
4. Six campers will get stars because that part is one half of the spinner.
5. 3

Page 66
[word search grid]

Page 69
1. Spider wanted to catch Snake.
2. Tiger told Spider he had to trick Snake.
3. Because he asked if Snake was longer than a bamboo pole.
4. Snake might say no.
5. Make-believe. The animals in the story talk.

Page 70
1. Mark has a dog.
2. His dog's name is Tiger.
3. Tiger can do tricks.
4. Mark throws a small ball.
5. Tiger jumps into the air and catches the ball.

Page 71
1. job
2. hear
3. trick
4. story
5. funny
6. house
7. grade
8. funny
9. story
10. trick
11. think

Page 72
1. 4
2. 8
3. 12
4. 14
5. 19
6. 19
7. 20
8. 25
9. 42
10. 51
11. 35
12. 64

Page 73
1. 13 pounds
2. 9 pounds
3. Possible answers: radio and tools; books and roller skates, books and radio; roller skates and radio
4. Yes. The total weight of the gifts is 32 pounds.

Page 74
1. flat hat
2. trick stick
3. grape cape
4. loud crowd

Page 77
1. She thought he would be good to eat.
2. Possible answers: the lady, cow, horse, and dog.
3. The gingerbread man got on the fox's back to cross the river.
4. The story says the fox wanted to eat the gingerbread man.
5. Answers will vary. Accept reasonable responses.

Page 78
1. What are you baking?
2. Can I lick the spoon?
3. What time will we eat?
4. May I set the table?
5. Would you pass the peas, please?

Page 79
1. to clean with water
2. to put together
3. to cook in an oven
4. one more than three
5. need
6. Four
7. cup
8. cut
9. Add
10. Bake

Page 80
1. 3
2. 4
3. 3
4. 9
5. 9
6. 8
7. 12
8. 17
9. 16
10. 21

Page 81
1. yes
2. yes
3. no
4–6. Answers will vary, but should show lines of symmetry.
7–9. Answers will vary, but should not show lines of symmetry.

Page 82
1. apple
2. cookie
3. carrot
4. oven
Riddle: pop corn

Page 84
1. D
2. C
3. B
4. A

Page 85
5. A
6. C
7. D
8. B

Page 86
9. B
10. C
11. D
12. A

Page 87
13. C
14. A
15. C
16. A

Page 88
1. B
2. C
3. C
4. B
5. A
6. D
7. D
8. A

Page 89
9. D
10. A
11. A
12. D
13. C
14. B
15. B
16. D
17. D

Page 90
18. C
19. D
20. B
21. C
22. C
23. A

96 Answer Key *Grade 2*